This book belongs to:

Tommy Garry

This 1989 Muppet Press book is published by Longmeadow Press.
Distributed by Checkerboard Press, a division of Macmillan, Inc.

Printed in U.S.A.
ISBN 022-689372-X
h g f e d c b a

Muppet Babies
Big Enough

by Harry Ross   illustrated by Kathy Spahr

Muppet Press

I'm big enough…
To brush my teeth,

To comb my hair

And wash my face.

I'm big enough…
To get my breakfast,

Find my spoon,
And set my place.

I'm big enough...
To choose my outfit
And not make
A great big mess.

I'm big enough…
To put my clothes on.

Isn't this
A pretty dress?

I'm big enough...
To get my shoes on

If they have
The straps that stick.

I'm big enough…
To put my coat on

If I use
This special trick.

I'm big enough...
To use the sandbox,

And I never
Throw the sand.

I'm big enough…
To swing myself.

I've learned to pump!
It's really grand.

I'm big enough…
To ride around,
Although it's only
On a trike.

But someday I'll be
Really cool
And ride all over
On my bike.

I'm big enough…
To print my name.
It's P-I-G-G-Y—
That's me!

I'm big enough…
To say my numbers.
Want to hear me?
One two three!

I'm big enough…
To do a puzzle.
I like
Dinosaurs the best.

I'm big enough…
To clean my room up—

Clothes in hamper,

Toys in chest.

And though there still are
Lots of things
That I can't do
'Cause I'm too small,

I'm getting bigger
All the time—
And someday I
Will do them all!